PROJECT
GEMINI

Pocket Space Guide

by Steve Whitfield

Gemini 9A astronauts, Eugene Cernan (left), and Thomas Stafford, receive a warm welcome as they arrive aboard the prime recovery ship, the aircraft carrier U.S.S. Wasp.

We acknowledge the financial support of the Government of Canada through the Book Publishing Industry Development Program for our publishing activities.

Published by Apogee Books, Box 62034, Burlington, Ontario, Canada, L7R 4K2, http://www.apogeebooks.com
Tel: (905) 637-5737

Printed and bound in Canada

Project Gemini Pcoket Space Guide by Steve Whitfield
Apogee Books Pocket Space Guide #12
ISBN-13: 978-1-894959-54-4
©2007 Robert Godwin

Contents

INTRODUCTION

Project Gemini, the second U.S. manned space program, was announced in January 1962. The third constellation of the Astrological Zodiac, Gemini, contains twin stars, Castor and Pollux, and a Gemini spacecraft has two astronauts, thus the name. Gemini missions (and their spacecraft) are properly named using Roman not Latin numerals—Gemini IV not Gemini 4—but both forms have been used extensively and this guide sticks with the simpler Latin numerals.

Project Gemini included 12 flights, two unmanned flight tests of the spacecraft and Titan-II launch vehicle, followed by 10 highly successful manned missions. Like Project Mercury, Project Gemini had major objectives that were clearly stated:

To subject men and equipment to space flight of up to two weeks duration.

To rendezvous and dock with orbiting vehicles and to maneuver the docked combination by using the target vehicle's propulsion system.

To perfect methods of entering the atmosphere and landing at a preselected point on land.

Project Gemini's goals were all achieved, with the exception of a landing on land, which was canceled in 1964 in favor of an ocean splashdown. Each Gemini flight also performed planned onboard (scientific and medical) experiments.

Another way of looking at Project Gemini is to consider it as the major learning curve in the American space program. Mercury was basically about getting into orbit, flying freefall, and then landing. Some maneuvering and experimenting were done, but it was basically the all-important baby steps. By the time of Apollo, the Mercury accomplishments were taken for granted. Apollo mission plans assumed that men could get to space, rendezvous and dock, stay healthy and sane for two weeks, etc.—all of the things that Project Gemini was spent learning and perfecting. In a 1999 interview veteran astronaut Jim Lovell said, "… we wanted to see how man could last for two weeks in space before we committed to go to the Moon …" and "… without Gemini, I don't think that Apollo would have been the success that it was."

Gemini is often lost in the shadows of Mercury (the first) and Apollo (the Moon), but many of its aspects were actually much more demanding then either other program. Launching atop a Titan-II booster was a more intense ride than with either the Mercury Atlas or the Apollo Saturn (and a space shuttle launch is a walk in the park by comparison). Neil Armstrong's life-saving recovery of the out-of-control Gemini 8 spacecraft was more difficult (and more miraculous) than anything done during Apollo, including the lunar landing. By the time of Apollo, all planned mission activities could be trained for and rehearsed ahead of time with a great deal of confidence, but only because that wasn't true for Gemini, where trial and error taught all of the important lessons. In short, if Mercury was about "the Right Stuff," Gemini was all about "the Essential Stuff."

GEMINI ASTRONAUTS

The crews of the Project Gemini missions included astronauts from the first, second and third NASA astronaut groups. Gus Grissom, Gordon Cooper and Walter Schirra were from the original "Mercury 7." John Young, James McDivitt, Edward White, Charles (Pete) Conrad, Thomas Stafford, Frank Borman, James Lovell and Neil Armstrong were from Group 2. David Scott, Eugene Cernan, Michael Collins, Richard Gordon and Buzz Aldrin were of the third astronaut group.

Astronaut Groups 1 and 2. The original seven Mercury astronauts selected by NASA in April 1959, are seated (left to right): L. Gordon Cooper Jr., Virgil I. Grissom, M. Scott Carpenter, Water M. Schirra Jr., John H. Glenn Jr., Alan B. Shepard Jr., and Donald K. Slayton. The second group of NASA astronauts, which were named in September, 1962, are standing (left to right): Edward H. White II, James A. McDivitt, John W. Young, Elliott M. See Jr., Charles Conrad Jr., Frank Borman, Neil A. Armstrong, Thomas P. Stafford, and James A. Lovell Jr.

MANNED FLIGHT SUMMARY

Gemini-Titan 3
Date: March 23, 1965   *Orbits*: 3
Duration: 0 days, 4 hrs, 52 min, 31 sec.
Astronauts: Gus Grissom, John Young

Gemini-Titan 4
Date: June 3-7, 1965   *Orbits*: 62
Duration: 4 days, 1 hr, 56 min, 12 sec.
Astronauts: James McDivitt, Edward White

Gemini-Titan 5
Date: August 21-29, 1965   *Orbits*: 120
Duration: 7 days, 22 hr, 55 min, 14 sec.
Astronauts: Gordon Cooper, Charles (Pete) Conrad

Gemini-Titan 6A
Date: December 15-16, 1965   *Orbits*: 16
Duration: 1 days, 1 hr, 51 min, 24 sec.
Astronauts: Walter Schirra, Thomas Stafford

Gemini-Titan 7
Date: December 4-18, 1965   *Orbits*: 206
Duration: 13 days, 18 hr, 35 min, 1 sec.
Astronauts: Frank Borman, James Lovell

Gemini-Titan 8
Date: March 16, 1966   *Orbits*: 7
Duration: 0 days, 10 hr, 41 min, 26 sec.
Astronauts: Neil Armstrong, David Scott

Gemini-Titan 9A
Date: June 3-6, 1966   *Orbits*: 47
Duration: 3 days, 0 hr, 20 min, 50 sec.
Astronauts: Thomas Stafford, Eugene Cernan

Gemini-Titan 10
Date: July 18-21, 1966   *Orbits*: 43
Duration: 2 days, 22 hr, 46 min, 39 sec.
Astronauts: John Young, Michael Collins

Gemini-Titan 11
Date: September 12-15, 1966   *Orbits*: 44
Duration: 2 days, 23 hr, 17 min, 8 sec.
Astronauts: Charles (Pete) Conrad, Richard Gordon

Gemini-Titan 12
Date: November 11-15, 1966   *Orbits*: 59
Duration: 3 days, 22 hr, 34 min, 31 sec.
Astronauts: James Lovell, Edwin (Buzz) Aldrin

SPACECRAFT DETAILS

The Gemini spacecraft was conical, 5.6 meters long, 3 meters in diameter at its base, and 1 meter in diameter at the top. Its two major sections were the Reentry Module and the Adapter Section.

Reentry Module

The Reentry Module was 3.35 meters high and 2.3 meters in diameter at its base. It had three main sections: (1) Rendezvous and Recovery (R&R), (2) Reentry Control Section (RCS), and (3) the cabin. The Rendezvous and Recovery section was the forward (small) end of the spacecraft, containing drogue, pilot and main parachutes and the radar.

The Reentry Control Section was between the R&R and cabin sections and contained fuel and oxidizer tanks, valves, tubing and two rings of eight attitude control thrusters for control during reentry. A parachute adapter assembly was included for main parachute attachment.

The cabin section housed the crew, seated side-by-side, and their instruments and controls. Above each seat was a hatch.

Adapter Section

The Adapter Section was 2.3 meters high and 3 meters in diameter at its base. It contained the Retrograde and Equipment sections.

The Retrograde section contained four solid retrograde rockets and part of the radiator for the cooling system.

The Equipment section contained fuel cells for electrical power, fuel for the Orbit Attitude and Maneuver System (OAMS), primary oxygen for the Environmental Control System (ECS), and cryogenic oxygen and hydrogen for fuel cell system. It also served as a radiator for the cooling system, also contained in the equipment section.

The Equipment section was jettisoned immediately before retrorockets were fired for reentry. The Retrograde section was jettisoned after the retros were fired.

Gemini Launch Vehicle

The Gemini Launch Vehicle (GLV) was a modified U.S. Air Force Titan-II intercontinental ballistic missile consisting of two stages. The overall height of launch vehicle and spacecraft was 33.2 meters. Their combined weight was about 154,220 kilograms.

Modifications to Titan-II for use as the Gemini Launch Vehicle included:

1. Malfunction detection system added to detect and transmit booster performance information to the crew.
2. Backup flight control system added to provide a secondary system if primary system fails.
3. Radio guidance substituted for inertial guidance.
4. Retro and vernier rockets deleted.
5. New second stage equipment truss added.
6. New second stage forward oxidizer skirt assembly added.
7. Trajectory tracking requirements simplified.
8. Electrical hydraulic and instrument systems modified.

Agena Target Vehicle

The Agena target vehicle was a modification of the U.S. Air Force Agena D upper stage, similar to the space vehicles which helped propelled *Ranger* and *Mariner* spacecraft to the Moon and planets. It acted as a separate stage of the Atlas / Agena launch vehicle, placing itself into orbit with its main propulsion system, and could be maneuvered either by ground control or the Gemini crew. Attitude control (roll, pitch, yaw) was accomplished by six nitrogen jets mounted on the Agena's aft end.

Atlas Launch Vehicle

The Atlas Standard Launch Vehicle was a refinement of the modified U.S. Air Force Atlas intercontinental ballistic missile, similar to the launch vehicle that placed the Project Mercury astronauts into orbit. It was used to launch the Agena target vehicles.

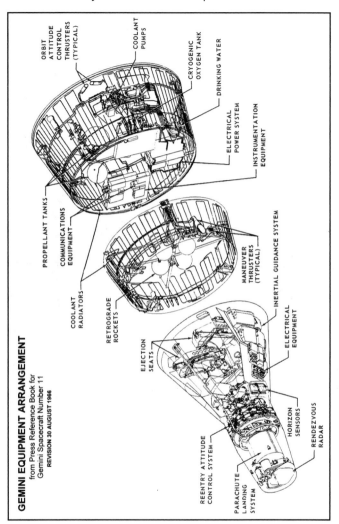

GEMINI EQUIPMENT ARRANGEMENT
from Press Reference Book for
Gemini Spacecraft Number 11
REVISION 30 AUGUST 1966

The Gemini spacecraft was conical, 5.6 meters long, 3 meters in diameter at its base, and 1 meter in diameter at the top. Its two major sections were the Reentry Module (bottom) and the Adapter Section.

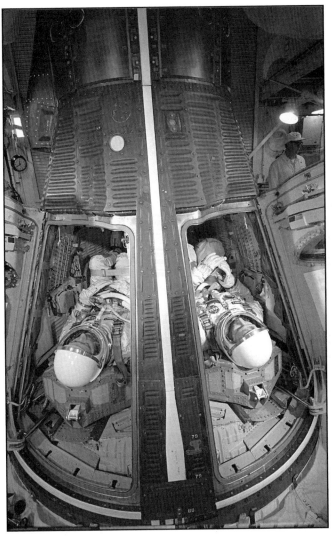

Gemini 4 astronauts, James McDivitt and Ed White, inside the Gemini spacecraft for a simulated launch at Cape Canaveral, Florida.

Gemini I (GT-1)

Orbits:	64
Launch:	April 8, 1964, 11:00:01.69 a.m. EST
Landing:	April 12, 1964
Crew:	Unmanned
Duration:	3 days, 23 hrs
Distance:	2,789,864 km
Altitude:	320 km x 160.3 km

The main objectives of this first flight in the program were to test the structural integrity of the Gemini spacecraft and the Gemini launch vehicle (GLV), a modified Titan-II ICBM. Specifically, the GLV performance and operation were being tested, along with its ground guidance systems. Additionally, the structural integrity and exit heating of GLV-spacecraft combination was being tested. A number of secondary objectives related to launch control, tracking and orbital insertion, were also included. Spacecraft #1 was built specifically for this mission and did not include life support systems. It was intended that the spacecraft would burn up on reentry.

The planned mission was only three orbits; the spacecraft stayed attached to the GLV second stage and there were no plans for recovery. The spacecraft remained in orbit for more than three days after the third orbit, then reentered the atmosphere and burned up.

The Titan-II GLV and Gemini spacecraft performed as required, both separately and as a combined flight unit. All primary and secondary objectives were achieved and the mission was termed a complete success.

Gemini II (GT-2)

Orbits:	Suborbital
Launch:	January 19, 1965, 9:03:59 a.m. EST
Landing:	January 19, 1965, 9:22:14 a.m. EST.
Crew:	Unmanned
Duration:	18 min, 16 sec
Distance:	3,422.4 km
Altitude:	171.2 km

Gemini 2 was a suborbital flight primarily for the purpose of verifying spacecraft performance and integrity from launch through reentry. An onboard automatic sequencer fired the retrorockets at the proper time to initiate the planned reentry. The landing was short of the planned splashdown point, but otherwise the flight path was as planned. The flight demonstrated satisfactory performance of: reentry heat protection during maximum heating; structural integrity of the spacecraft; major spacecraft subsystems; and backup guidance steering signals. Secondary mission objectives related to additional spacecraft and GLV tests and training of flight controllers.

The Gemini 2 launch vehicle and spacecraft received much attention prior to launch day. They were dismantled twice in the summer of 1964 to protect them from hurricanes. They were also used for ground control tests, astronaut and ground crew training, and full-scale launch countdown rehearsals for manned missions.

On December 9, 1964, the originally scheduled launch date, the countdown actually reached engine ignition, only to be aborted by the launch vehicle's Malfunction Detection System one second later because of a loss of hydraulic pressure. When it was finally launched on January 19, 1965, everything went smoothly. All primary and secondary mission objectives were achieved, with the exception of the fuel cell test; the fuel cell malfunctioned and shut down prior to liftoff.

Gemini III (GT-3)

Nickname: *Molly Brown*
Orbits: 3
Launch: March 23, 1965
Landing: March 23, 1965
Commander: Virgil I. (Gus) Grissom
Pilot: John W. Young
Duration: 4 hours, 52 minutes, 31 seconds
Distance: 128,748 km
Altitude: 224 km
Inclination: 32.6 degrees
Backup crew: Walter M. Schirra, Jr., Thomas P. Stafford
CapComs: Gordon Cooper (Cape);
               Roger Chaffee (Houston)

Gemini 3 was the first manned Gemini flight, and the first two-man American space flight. It flew three orbits, primarily to evaluate the two-man spacecraft design. This was the only Gemini spacecraft to be given a nickname, as had been done with the Mercury spacecraft. Grissom named it *Molly Brown*, a reference to the Broadway show, *The Unsinkable Molly Brown*, because his Mercury spacecraft, *Liberty Bell 7*, did in fact sink after splashing down, and Grissom came close to sinking himself.

Mission Objectives

This mission was to test and verify the Gemini spacecraft, which provided the pilot with maneuverability that had not been available in the Mercury spacecraft through a system called OAMS (Orbit Attitude and Maneuvering System). OAMS could be used to manually adjust the spacecraft's orbit (instead of just its attitude) and, in addition, could produce lift by varying the spacecraft's angle of attack during reentry, which gave some control over where the spacecraft would land. (Angle of attack is the difference between the spacecraft's direction of travel and the direction that it is pointing; think of an airplane coming *down* for a landing; its nose is actually pointing *up*.)

In addition to the OAMS testing, as always, a wide variety of spacecraft and GLV systems was being tested from launch through splashdown. Some general photography from orbit was also done.

Launch
Gemini 3 was launched on March 23, 1965 at 9:24:00 a.m. EST with no problems. There was a single brief hold during countdown for a sensor adjustment.

Mission Highlights
All primary objectives were achieved except the controlled reentry, which was partially achieved (the angle of attack during reentry was lower than expected). The secondary mission objectives were only partially achieved. One zero-gravity experiment had an equipment error and the orbital photography was only partially successful.

Landing
Gemini 3 splashed down in the Atlantic Ocean, near Grand Turk Island in the West Indies, on March 23, 1965 after three orbits, approximately 111 km from the planned impact point.

View of the Flight Director's console in the Mission Control Center (MCC), Houston, Texas, during the Gemini 5 flight.

Gemini IV (GT-4)

Orbits:	62
Launch:	June 3, 1965
Landing:	June 7, 1965
Commander:	James A. McDivitt
Pilot:	Edward H. White II
Duration:	4 days, 1 hr, 56 min, 12 sec.
Distance:	2,590,561 km
Altitude:	296.1 km
Inclination:	32.5 degrees
Backup crew:	Frank Borman, James A. Lovell, Jr.
CapComs:	Clifton C. Williams, Jr. (Cape);
	Gus Grissom (Houston)

Gemini 4 was the 10th American manned space flight and it included the first extravehicular activity (EVA) by an American. Ed White performed a 22-minute "space walk" with no problems (in fact, he enjoyed it so much that he had to be told twice to return to the capsule). The EVA was not originally planned as part of the mission, but the USSR had recently performed a space walk it was felt necessary to match them, rather than keep falling farther behind.

Mission Objectives

This four-day mission was America's first multiple-day space flight. If men were to plan a Moon landing, it first had to be determined whether it was possible for a man to function and remain healthy in space long enough to get there, land and explore, and come back to Earth. Gemini 4 was set up as an evaluation of procedures for crew rest and work cycles, eating schedules, and real-time flight planning.

Secondary objectives included: tethered EVA (space walk); rendezvous and stationkeeping with the GLV second stage; further evaluate of spacecraft systems; and orbital plane change maneuvers. There were also 11 onboard experiments to be performed.

Launch

After an uneventful countdown, launch occurred on June 3, 1965 at 10:15:59 a.m. EST. This was the first spacecraft launch to be televised live internationally.

Mission Highlights

Gemini 4 pilot Ed White spent 22 minutes performing a space walk, tied to a tether and using a handheld "zip gun" to maneuver himself around the spacecraft, while Commander Jim McDivitt took photographs. Gemini 4 set a new record for flight duration, and eased fears about the medical consequences of longer missions. It was also the first use of the new Mission Control Center near Houston, Texas.

One of the secondary objectives for this mission was for Gemini 4 to fly in formation with the spent second stage of its Titan-II launch vehicle while in orbit. Their attempts failed completely, so they quit trying, since their fuel was being seriously depleted. When thrusting toward their target, they only managed to move farther away from it. They learned the hard way a basic rule of orbital mechanics that should have been foreseen—in a free-fall orbit, thrusting in the direction of travel doesn't make you go faster in the same orbit, it puts you into a higher orbit, which effectively makes you go slower with respect to the target you were pointing at! This is completely counterintuitive to people who have lived their entire lives on the surface of a planet. On later missions rendezvous was successfully completed by approaching the target from below in a lower, faster orbit and then thrusting up to catch it.

All other secondary objectives were met and the only primary objective not achieved was computer-controlled reentry, which could not be executed because of inadvertent alteration of computer memory.

Landing

Gemini 4 splashed down in the Atlantic Ocean on June 7, about 81½ km from its planned landing zone 540 km south of Bermuda.

Gemini 4 astronaut Edward H. White II is shown in the crew ready room at Launch Complex 16, suited and ready to ride the van to Launch Complex 19 for insertion in the spacecraft. White successfully accomplished the first U.S. spacewalk during the Gemini 4 mission.

Gemini V (GT-5)

Orbits:	120
Launch:	August 21, 1965
Landing:	August 29, 1965
Commander:	L. Gordon Cooper
Pilot:	Charles (Pete) Conrad, Jr.
Duration:	7 days, 22 hr, 55 min, 14 sec.
Distance:	5,242,682 km
Altitude:	349.8 km
Inclination:	32.61 degrees
Backup crew:	Neil A. Armstrong, Elliott M. See, Jr.
CapComs:	Gus Grissom (Cape); Jim McDivitt, Buzz Aldrin & Neil Armstrong (Houston)

Gemini 5 doubled the space flight duration record of Gemini 4 with this eight-day mission. This is about the length of time that it would take to fly a mission to the Moon. This was the second time that Gordon Cooper set a space flight duration record (his Mercury flight was more than 34 hours). Gemini 5 was the first spacecraft to use a fuel cell for electrical power.

Mission Objectives

The primary objectives of Gemini 5 were to demonstrate 8-day capability of the spacecraft and crew; evaluate the effects of weightlessness during an 8-day flight; and evaluate the rendezvous Guidance and Navigation System using a rendezvous pod ejected from the spacecraft.

The secondary objectives included demonstrating controlled reentry guidance; evaluating the fuel cell; and demonstrating all phases of Guidance and Control System operation needed for rendezvous. In addition, both crewmen were to maneuver spacecraft to rendezvous, check out the rendezvous radar, and execute 17 onboard experiments.

Launch

Gemini 5 was launched August 21, 1965 at 8:59:59 a.m. EST.

Launch was attempted on August 19, but was postponed because of weather conditions and problems with loading cryogenic fuel for the fuel cell.

Mission Highlights

At two hours and 13 minutes into the flight the rendezvous pod was ejected and the rendezvous radar showed that the pod was moving at a speed of two m/sec relative to the spacecraft. After four hours and 22 minutes they found that the pressure in their fuel cell had dropped very significantly. Cooper decided to shut it down, but without electrical power they would be unable to rendezvous with the pod, which could mean ending the mission prematurely. With the fuel cell off, they could only stay in orbit for one day and still have enough battery power for reentry. Testing showed that the fuel cell would work, even with the low oxygen pressure, so it was turned back on and this was confirmed, so the mission continued.

On the ground, Buzz Aldrin had worked out an alternative rendezvous test that didn't require the rendezvous pod, which had been lost during electrical power-down.

Cooper and Conrad had trouble sleeping during alternating sleep periods and eventually decided to take their sleep periods together. Gemini 4 had had the same problem. In their semi-powered drifting flight the crew found that seeing stars slowly drifting past the windows was disorienting, so they covered over the windows.

The revised rendezvous exercise was performed on day 3 without any problems. It was the first-ever precision orbital maneuver done on a US space flight. In all, they performed four OAMS maneuvers—apogee adjust, phase adjust, plane change, and a coelliptical maneuver (matching orbits).

On day 5, one of OAMS thrusters stopped working, and they couldn't resolve the problem, so all experiments requiring fuel were canceled. However, most of the 17 planned onboard experiments were carried out, the majority of which were related to visual and photographic studies.

Attempts to manually control reentry were successful; drag and lift were created by using the OAMS to alter the spacecraft's attitude. Because of a programmer data error in the onboard computer, Gemini 5 landed well short of their planned impact point.

Landing

Gemini 5 splashed down in the Atlantic Ocean on August 29, 1965, about 170 km from their intended impact point. The backup recovery team from the USS DuPont transferred them via helicopter to the prime recovery vessel, USS Lake Champlain, within 1½ hours.

Gemini 5 astronauts Gordon Cooper (right) and Charles "Pete" Conrad walk across the deck of the recovery aircraft carrier U.S.S. Lake Champlain following splashdown and recovery from the ocean.

Gemini VII (GT-7)

Orbits: 206
Launch: December 4, 1965
Landing: December 18, 1965
Commander: Frank Borman
Pilot: James A. Lovell
Duration: 13 days, 18 hr, 35 min, 1 sec.
Distance: 9,029,771 km
Altitude: 327 km
Inclination: 28.89 degrees
Backup crew: Edward H. White II, Michael Collins
CapComs: Alan Bean (Cape); Elliott See,
               Eugene Cernan & Charles Bassett (Houston)

Gemini 7's most significant accomplishment was not part of its original mission plan. When the Gemini 6 mission was scrubbed because its Agena rendezvous target vehicle had failed, the Gemini 7 spacecraft was used for the rendezvous in its place. Gemini 7 was the 12th manned American flight and set a new long-duration US space mission record that lasted for five years (until Skylab).

Mission Objectives

The primary mission objective for Gemini 7 was to conduct a 14-day mission and evaluate effects on crew health and performance. Included in the mission plan were attendant problems such as timing their activities to match the work day of the ground crews.

Originally intended to fly after Gemini 6, Gemini 7 was timed such that Gemini 6 could be launched and rendezvous with them in orbit. Gemini 6's original flight plan called for rendezvous and docking with an Agena target vehicle; with Gemini 7 it could only rendezvous and stationkeep. there was no complementary docking port on the Gemini 7 spacecraft.

Secondary Gemini 7 objectives included: stationkeeping with

its GLV second stage in orbit; conducting 20 onboard experiments (the most of any Gemini mission) including studies of nutrition in space; evaluate the new lightweight pressure suit; and evaluate spacecraft reentry capabilities.

Launch
Gemini 7 was launched December 4, 1965 at 2:30:03 p.m. EST, the first US manned flight not to be launched in the morning. Launch and ascent were without incident.

Mission Highlights
Once in orbit Gemini 7 spent 15 minutes stationkeeping with its GLV second stage, then Borman elected to discontinue because they were using too much fuel. The stage was itself venting fuel, which caused it to continue moving.

On day 2 Lovell was reluctantly given permission to remove his space suit. They were wearing and evaluating a new "soft suit" designed for long-duration missions. Both astronauts had been sweating continuously and Lovell had been slowly removing his suit in stages. NASA managers didn't approve and said that one crew member had to remain suited at all times. Their evaluation was that the lightweight space suit was uncomfortable if worn too long in Gemini's hot, cramped confines. After two days NASA relaxed the suit requirement.

Over a five-day period they had made four orbital adjustment burns and putting them in a 300-kilometer circular orbit, meaning that they could stay in orbit for at least 100 days without the orbit degrading. This is where Gemini 6's rendezvous radar found them on December 15. Gemini 6 was in a lower orbit, gaining on Gemini 7 with a series of manual burns. Once they were within radar lock Gemini 6's computer took over rendezvous maneuvers and closed the distance between the two spacecraft to 40 meters. During the fly-arounds and stationkeeping that followed the two spacecraft moved as close as 30 centimeters. Gemini 6 backed off for the sleep period and next "morning" flew off and reentered. Three days later Gemini 7 did likewise.

The Gemini VII Crew, James Lovell, pilot, and Frank Borman, command pilot, leave the suit trailer on the way to Pad 19. They are wearing the specially-designed lightweight suits.

During the final three days of the mission, following the rendezvous exercises, Borman and Lovell spent time reading; Pete Conrad, based on his Gemini 5 experience, had suggested that they take books along to relieve the boredom. The Gemini 7 astronauts were somewhat weakened by their two weeks in space, but both were healthy in body and mind. All primary and secondary mission objectives were achieved with only very minor spacecraft malfunctions.

Landing
Gemini 7 splashed down in the west Atlantic Ocean on December 18, 1965, only 11.8 km from its planned impact point.

Gemini VI (GT-6A)

Orbits: 16
Launch: December 15, 1965
Landing: December 16, 1965
Commander: Walter M. Schirra Jr.
Pilot: Thomas P. Stafford
Duration: 1 days, 1 hr, 51 min, 24 sec.
Distance: 694,415 km
Altitude: 311.3 km
Inclination: 28.89 degrees
Backup crew: Virgil I. Grissom, John W. Young
CapComs: Alan Bean (Cape); Elliott See, Eugene Cernan &
               Charles Bassett (Houston)

Gemini 6 was to be the first spacecraft to rendezvous and dock with an Agena target vehicle. Its Agena failed to reach orbit, so Gemini 6A later rendezvoused in orbit with the Gemini 7 spacecraft instead.

Mission Objectives

Gemini 6's primary objective was to rendezvous and dock with an Agena target vehicle. When the Agena launch failed and the Agena was lost, the primary objective was revised to a rendezvous with the Gemini 7 spacecraft, and the mission was renamed Gemini 6A.

The secondary objectives included: rendezvous in the fourth orbit (rendezvous and docking early in the mission was a critical requirement for Apollo which had to be perfected during Gemini); stationkeeping with Gemini 7; evaluating reentry guidance capability; and conducting visibility tests for rendezvous using Gemini 7 as target. They also had to perform three experiments.

Launch

The Gemini 6 mission, set for October 25, 1965, was rescheduled to December 12 when its Agena target vehicle

(GATV) failed to reach orbit. The December 12 launch attempt failed because of a minor launch vehicle hardware problem. The Gemini 6A spacecraft was finally launched on December 15, 1965 at 8:37:26 a.m. EST on its way to rendezvous in orbit with Gemini 7.

Mission Highlights

Launch and power flight were good and the spacecraft went into a low orbit. The plan called for rendezvous on the fourth orbit. They made burns at 94 minutes and at 2 hours, 18 minutes after launch, which put them in the same orbital plane as Gemini 7 and trailing by 483 km. The rendezvous radar made contact with Gemini 7 at 3 hours, 15 minutes, at a range of 434 kilometers.

A third burn put them into an orbit below Gemini 7 and slowly gaining on them. At this point, Schirra let the Gemini 6A computer take over the rendezvous maneuvers. At 5 hours, 4 minutes in the flight Gemini 7 became visible as a bright star. After several more computer-controlled burns the two spacecraft were only 40 meters apart. During the next 4½ hours the two spacecraft were as close as 30 centimeters to 90 meters apart. Stationkeeping was entirely successful. Gemini 6A moved out to 16 kilometers from Gemini 7 for the sleep period, and following sleep they reentered and landed.

However, several hours before retrofire, during a pass over the States, Schirra made the following report to Mission Control: "This is Gemini 6. We have an object, looks like a satellite, going from north to south, up in a polar orbit. He's in a very low trajectory … looks like he may be going to reenter pretty soon. Stand by … it looks like he's trying to signal us." This transmission was immediately followed by "Jingle Bells," played by harmonica and bells. Thus, the spirit of the season was brought into the mission. Speaking of that report, at the Gemini 7-6 news conference in Houston on December 30, Schirra said, in part, "… Our intent was not a prank. It was to relieve the tension … I think we convinced

Chris [Kraft, Flight Director] and many of the people on the flight control team that we did, in fact, have an unidentified flying object there. And, I think the children of this country are happier for the fact that we might have seen something there."

All primary mission objectives and all but one of the secondary objectives were achieved.

Landing

Gemini 6A landed in the west Atlantic Ocean on December 16, 1965, 12.9 km from its planned impact point. This was the first landing to be televised live.

An artist's concept of the Gemini 6A and Gemini 7 spacecraft stationkeeping after rendezvousing in orbit. The scene depicted did actually take place on December 15-16, 1965.

Gemini VIII (GT-8)

Orbits: 7
Launch: March 16, 1966
Landing: March 16, 1966
Commander: Neil A. Armstrong
Pilot: David R. Scott
Duration: 0 days, 10 hr, 41 min, 26 sec.
Distance: 293,206 km
Altitude: 298.7 km
Inclination: 28.91 degrees
Backup crew: Charles (Pete) Conrad, Richard F. Gordon
CapComs: Walter Cunningham (Cape);
                James Lovell (Houston)

Gemini 8 accomplished America's first docking with another space vehicle, an unmanned Agena stage. A malfunction caused uncontrollable spinning of the spacecraft, so the crew undocked and effected the first-ever emergency landing of a manned U.S. space mission.

Mission Objectives

The primary Gemini 8 mission objectives were to rendezvous and dock with an Agena target vehicle and to conduct extended EVA (space walk) operations.

The mission's secondary objectives included: rendezvousing and docking in fourth orbit and performing docked-vehicle maneuvers. There were also 10 onboard experiments to conduct.

Launch

Gemini 8 was launched on March 16, 1966 at 11:41:02 a.m. EST after a one-day delay because of minor problems with the spacecraft and launch vehicle hardware.

Mission Highlights

This was the first time that NASA managed to successfully launch a Gemini spacecraft and its Agena target vehicle together. Gemini 8 performed the world's first orbital docking just 6½ hours after the Gemini launch—but half an hour later it all went wrong.

After its first orbit Gemini 8 made three burns to put it into the same plane as the Agena and into capture position. At this point a ground controller realized that they were slightly off because one or more thrusters was misbehaving, so they had to make an additional minor correction burn.

The rendezvous radar acquired the Agena at a range of 332 km. At 3 hours, 48 minutes mission time they burned again which put them 28 km below the Agena's orbit. They sighted the Agena at 140 km range and let the computer take over rendezvous maneuvers at 102 km. Armstrong piloted Gemini 8 to within less than a meter of the Agena and then slowly docked to it.

At about 30 minutes after having docked, the docked Gemini-Agena system began an uncommanded roll. Using the Gemini's OAMS, Armstrong got the roll stopped, but as soon as he disengaged the OAMS, the roll started up again. Suspecting that the problem was with the Agena, they immediately turned it off and the problem stopped. But a few minutes later it started up again.

They realized that it was a problem with the Gemini spacecraft when they saw that its attitude fuel was down to 30%.

They transferred control of the Agena back to the ground controllers and then undocked from it and moved away. Unfortunately, the Gemini spacecraft then began rolling even faster, reaching one revolution per second. At that point the only thing to do was turn off OAMS and use the Reentry Control System (RCS) to try to eliminate the roll, which by now had become dangerous to the astronauts; if it continued they would soon black out. Once they had stopped the roll using the RCS, they experimented with the OAMS thrusters and confirmed that one of them was stuck open.

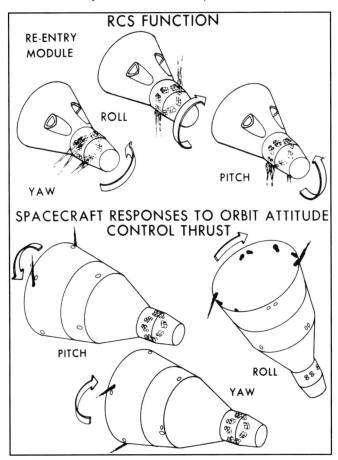

Positioning and effects of the Gemini spacecraft's Reentry Control System and Orbit Attitude and Control System.

Armstrong and Scott were forced to cut short their mission and make an emergency reentry one orbit later, 10 hours after launch, splashing down in a secondary landing area in the Pacific Ocean. Although the rendezvous and docking were a complete success, David Scott did not get to make his planned two-hour space walk.

Neil Armstrong, an experienced test pilot, including test flights of the X-15 rocket, put in an unparalleled performance in a dangerous and immediate situation.

Landing
Gemini 8 splashed down in the west Pacific Ocean on March 16 at 10:22:28 p.m. EST. Landing was planned for the Atlantic Ocean. It was hours before a backup recovery team had the astronauts onto the carrier.

Gemini IX (GT-9A)

Orbits:	47
Launch:	June 3, 1966
Landing:	June 6, 1966
Commander:	Thomas P. Stafford
Pilot:	Eugene A. Cernan
Duration:	3 days, 0 hr, 20 min, 50 sec.
Distance:	2,020,741 km
Altitude:	311.5 km
Inclination:	28.86 degrees
Backup crew:	James Lovell, Edwin (Buzz) Aldrin
CapComs:	Buzz Aldrin (Cape, Houston); Neil Armstrong, James Lovell & Richard Gordon (Houston)

Stafford and Cernan, originally the backup crew, were promoted to prime crew, when Elliott See and Charles Bassett died in a plane crash four months before the launch.

Mission Objectives
The primary objectives of the Gemini 9 mission were to rendezvous and dock with an Agena target vehicle, and to conduct extended EVA activities using an Air Force Astronaut Maneuvering Unit (AMU), predecessor of the Manned Maneuvering Unit (MMU) later used by Shuttle astronauts.

The secondary objectives included: rendezvousing during third orbit; practice docking and rendezvous from above (from a higher orbit); and demonstrate controlled reentry.

Launch

Gemini 9 was scheduled to launch on May 17, 1966 but was postponed when the Atlas booster for its Agena target vehicle malfunctioned. Launch was rescheduled to June 1 and an Augmented Target Docking Adapter (ATDA) was used in place of the Agena. The ATDA was launched on schedule but Gemini 9, renamed Gemini 9A, experienced a guidance system computer problem and was not launched until June 3.

Mission Highlights

Two burns made early in the mission put Gemini 9A where they wanted to be and closing on the ATDA at 38 m/sec. The rendezvous radar locked onto the ATDA at a range of 222 km. Telemetry had indicated that the ATDA's launch shroud had not been jettisoned, but at 93 km range they could see the flashing lights on the ATDA (installed to aid identification from a distance) which suggested that the shroud had in fact been jettisoned. Once they were within visual range they could see that the shroud had only partially detached and was hanging at an angle. Stafford said, "It looks like an angry alligator out here rotating around," and the name stuck; to this day, any pictures or text pertaining Gemini 9 include the "angry alligator."

Stafford suggested trying to use the Gemini spacecraft to "open the jaws," but permission was not granted. No safe way could be found to detach the shroud, so it was left in place. It was later determined that the problem had resulted from incomplete communication between the multiple contractors involved with the ATDA. This was not a unique situation in the US space program; a major later example was the voltage mismatch that caused the Apollo 13 explosion.

During the remainder of day 1 and day 2 they practiced various approach and rendezvous activities. While stationkeeping with the ATDA, they were given permission to start EVA, but they were tired and Stafford didn't want to continue stationkeeping (they couldn't dock because of the shroud) so they postponed EVA until day 3.

Cernan began his EVA activities on day 3 and it became a real eye-opener. On one hand, nothing went as planned; on the other hand, they learned a great deal about working in space—critical experience that was reinforced on subsequent EVA missions. Almost everything was harder than had been anticipated because of the lack of hand and foot holds, which made it impossible to gain any leverage for even the simplest of operations. This is something that should have been foreseen, and it's very surprising that it wasn't.

Additionally, when working in the dark Cernan's faceplate fogged up and his pulse was soaring (195 beats per minute), and the flight surgeon was concerned that he might lose consciousness. Considering this and the remaining planned EVA activities, he decided to cancel the rest of the EVA, a decision with which everyone involved agreed. He had spent more than two hours outside the spacecraft. He was so overheated that his faceplate fogged up completely before he was back through the hatch and Stafford had to help him in. It was a near thing.

So, both of the mission's primary objective (rendezvous and docking, and extended EVA) were only partially achieved. The Astronaut Maneuvering Unit did not get used, but Cernan set a new EVA duration record (2 hours, 9 minutes) and learned a great deal about what needed to be done to facilitate EVA operations. These requirements became a main focus of later mission plans, particularly Gemini 12.

Landing

Gemini 9A splashed down in the west Atlantic Ocean about 345 miles east of Cape Kennedy, less than 1 km from the planned impact point.

Gemini X (GT-10)

Orbits: 43
Launch: July 18, 1966
Landing: July 21, 1966
Commander: John W. Young
Pilot: Michael Collins
Duration: 2 days, 22 hr, 46 min, 39 sec.
Distance: 1,968,823 km
Altitude: 753.3 km
Inclination: 28.85 degrees
Backup crew: Alan L. Bean, Clifton C. Williams Jr.
CapComs: Gordon Cooper (Cape, Houston);
              Buzz Aldrin (Houston)

Gemini 10 was the first mission to use an Agena target vehicle's propulsion systems to control docked spacecraft. It was also the first spacecraft to rendezvous with two different target vehicles.

Mission Objectives
The primary objective of Gemini 10 was to rendezvous and dock with an Agena target vehicle launched the same day.

The secondary objectives included: docking during the fourth orbit; rendezvous with the Agena target vehicle from the Gemini 8 mission; conduct EVA exercises; and leave the Gemini 10 Agena target vehicle parked in a specific orbit.

Launch
The Agena launched successfully and Gemini 10 launched 100 minutes later at 5:20:26 a.m. EST on July 18, 1966. Their initial orbit put them 1,800 km behind the Agena.

Mission Highlights
The planned use of a sextant for navigation didn't work out as expected, so ground computer calculations were used instead during the mission.

After the first two orbital burns of the mission they discovered that their orbit had an out-of-plane error, so two additional corrections burns were required to bring them to where they needed to be to rendezvous with their Agena target vehicle. After acquiring and docking with the Agena the decision was made to remain docked as long as possible and use Agena fuel for attitude control.   Using the Agena propulsion system, controlled from the Gemini spacecraft, they performed a burn that put the docked pair into a 294 km by 763 km orbit. The 763 km apogee was the highest altitude that anyone had ever reached to that time. On day 2 they made a second burn with the Agena engine to put them into an orbit matching the drifting Agena from the abbreviated Gemini 8 mission. Another Agena burn followed to circularize their orbit at 377.6 km.

The next thing on their to-do list was the first of two EVAs, a stand-up EVA during which Collins would remain within the Gemini hatch opening to perform some planned star photography with a 70 mm general purpose camera and ultraviolet film. The EVA was terminated six minutes early because of eye irritation.

Also scheduled for day 2 was the second rendezvous of the Gemini 10 mission. Their own Agena, from which they had undocked previously, was relatively near by, but the Gemini 8 Agena, their next rendezvous target, was much farther away. They acquired it visually (as a faint star) once the range had been reduced to 30 km. Two more correction burns brought them into stationkeeping only three meters away from the Agena.

The start of day 3 in orbit was the time for the second planned EVA. With some difficulty, Collins retrieved a micrometeorite collector from the side of the Gemini spacecraft, which was later lost some time during the EVA when it must have floated out of the cabin. Leaving the Gemini spacecraft, he moved to the Agena target vehicle using the gas gun designed for the purpose (like Cernan before him, he was hampered by the lack of handholds). He retrieved a second micrometeorite collector from the Agena.

25 minutes into the EVA, the gas gun stopped working, ending EVA activities. The Gemini hatch was reopened a third time to jettison the EVA chestpack and umbilical cord.

Landing
Gemini 10 splashed down on July 21 at 4:07:05 p.m. in the west Atlantic Ocean about 5½ km from their planned impact point.

Gemini XI (GT-11)

Orbits:	44
Launch:	September 12, 1966
Landing:	September 15, 1966
Commander:	Charles (Pete) Conrad Jr.
Pilot:	Richard F. Gordon Jr.
Duration:	2 days, 23 hr, 17 min, 8 sec.
Distance:	1,983,565 km
Altitude:	1,368.9 km
Inclination:	28.83 degrees
Backup crew:	Neil Armstrong, William Anders
CapComs:	Clifton C. Williams Jr. (Cape);
	John Young & Alan Bean (Houston)

Gemini 11 set the record for the highest Earth orbit altitude by an American manned spacecraft (1,374 km). Its most important accomplishment was rendezvous and docking at the end of their first orbit, which would be required for Apollo lunar landings.

Mission Objectives
The primary objective of Gemini 11 was to rendezvous and dock with the Agena target vehicle at the end of their first orbit. After much debate, the upcoming Apollo lunar missions being planned would use Lunar Orbit Rendezvous (LOR), which meant that the lunar lander would separate from the command module in lunar orbit and land on the Moon. At the

end of the lunar excursion the lander would lift off from the surface and rendezvous with the command module in lunar orbit and they would dock. To make LOR work, rendezvous and docking had to be accomplished at the end of the lander's first orbit after ascent—there was no fallback position; it had to go right on the first try. Gemini 11 was the first trial run at this, to see if the machines, and more importantly the astronauts, were capable of doing it.

The mission's secondary objectives included: practicing docking; two planned EVAs; maneuvering while docked; a tethered vehicle test; and automatic computer-controlled reentry.

Launch

The Gemini 11 launch was postponed on September 9 and again on September 10 because of launch vehicle malfunctions. It was finally launched on September 12, 1966 at 9:42:26 a.m. EST. The Agena target vehicle had been launched just under an hour earlier.

Mission Highlights

Gemini 11 started off with a major high—rendezvous and docking in the first orbit was the main highlight of the mission. They undocked and redocked several times, and then remained docked to the Agena and used its propulsion system to move into a record high manned spacecraft altitude of nearly 1,400 km.

Towards the end of the first day of the flight, Richard Gordon began the first of two planned EVAs, this one involving both the Gemini and Agena spacecraft. As had been found on Gemini 9A and Gemini 10, the EVA was much more difficult and physically demanding than ground-based training had led them to expect. Gordon became overtired and cut the EVA short.

At the end of day 2 was the second EVA, this time a "stand-up" EVA, performed within the Gemini spacecraft hatch opening where the astronaut could be anchored to the spacecraft by his partner when needed. Again consistent with earlier flights, the stand-up EVA was much easier and proceeded more or less according to plan.

Gemini 11 also performed a tether experiment, the first time that this had been attempted. During the first EVA, Gordon had attached a tether to both spacecraft. The two vehicles were now undocked and the Gemini was maneuvered so as to pull the tether taut. The theory was that rotation of the tethered spacecraft (around their mutual center of gravity) would cause a slight "artificial gravity" to be induced in both vehicles. The experiment was difficult to perform because Conrad had problems keeping the tether taut, but indications were that they succeeded to some degree.

The final highlight of the mission—a significant one in terms of upcoming Apollo requirements—was the first completely automatic, computer-controlled reentry. The system worked extremely well, bringing Gemini 11 down only 4.5 kilometers from its recovery ship.

Landing
Gemini 11 splashed down in the Atlantic Ocean on September 15, 1966, less than 5 km from its planned impact point.

Gemini XII (GT-12)

Orbits:	59
Launch:	November 11, 1966
Landing:	November 15, 1966
Commander:	James A. Lovell, Jr.
Pilot:	Edwin E. (Buzz) Aldrin, Jr.
Duration:	3 days, 22 hr, 34 min, 31 sec.
Distance:	2,574,950 km
Altitude:	270.6 km
Inclination:	28.87 degrees
Backup crew:	Gordon Cooper, Eugene Cernan

CapComs: Stuart Roosa (Cape); Pete Conrad & William
              Anders (Houston)

Gemini 12 was the 10th and final manned Gemini flight. As a
result of the problems on earlier Gemini missions, Gemini 12
concentrated on improved EVA procedures and facilities. It
was the first mission to conduct three EVAs.

Mission Objectives

The primary objective of the Gemini 12 mission was to
rendezvous and dock with an Agena target vehicle and use the
docked vehicles as a test platform for evaluating EVA procedures.

The secondary objectives included: tethered vehicle
stationkeeping; practice docking and docked maneuvers; park the
Agena target vehicle in a specified orbit; and automatic reentry.

Launch

Gemini 12 was launched on November 11, 1966 at 3:46:33 p.m.
EST, about 1½ hours after its Agena target vehicle was launched.

Mission Highlights

At about four hours into this four-day mission Lovell and
Aldrin rendezvoused and docked with their Agena target
vehicle. A few hours later they used the Agena propulsion
system to burn the docked spacecraft into a higher orbit.

In the later hours of day 1 Aldrin performed his first planned
EVA, a "stand-up" EVA which lasted 2 hours and 29 minutes.
While in daylight, the crew practiced umbilical EVA
procedures, and then Aldrin installed a telescoping handrail
between the two spacecraft. It would afford Aldrin a handhold
during his later umbilical EVA movement to the Agena. During
this stand-up EVA Aldrin was connected to the spacecraft by
three umbilicals (providing environmental control system and
electrical/comm connections) and a nylon-webbing tether.

During day 2, after another Agena propulsion burn had moved
them to a lower orbit, Aldrin performed a two-hour and six-
minute "umbilical" EVA. One activity during this EVA was to

attach the Agena tether to be used for the tethered stationkeeping experiment. Throughout the EVA he performed diverse tasks using various foot restraints and/or body tethers to see how they helped to maintain his body position. The EVA included planned rest periods.

During day 3, Aldrin performed his second stand-up EVA, which lasted 55 minutes. This EVA consisted primarily of jettisoning unnecessary EVA equipment and performing planned photography tasks. This third EVA brought his total EVA time for the mission to 5½ hours, a very significant improvement over previous missions.

The mission concluded with an entirely successful autopilot reentry. All in all, Gemini 12 was a fine way to finish the Gemini program.

Landing

Gemini 12 splashed down in the Atlantic Ocean on November 15, 1966, less than 5 km from its planned impact point.

SUMMARY OF ACCOMPLISHMENTS

All Project Gemini objectives, including EVA and docked vehicle maneuvers (added after project start), were fully accomplished many times over, as listed below:

Rendezvous: Ten separate rendezvous were accomplished, using seven different techniques ranging from visual / manual control to ground / computer controlled rendezvous.

Docking: Nine dockings with four different Agenas were performed.

Docked Vehicle Maneuvers: Both Gemini 10 and Gemini 11 demonstrated extensive maneuvers and a new altitude record was set on Gemini 11 when the Agena target vehicle carried astronauts Conrad and Gordon 1,374 km above the Earth.

Extravehicular Activity: EVA was conducted on five separate Gemini missions, during ten separate periods. The total EVA time for Project Gemini was 12 hours and 22 minutes, of which a record time of 5½ hours was performed by Buzz Aldrin on Gemini 12.

Long Duration Flight: Gemini 7 demonstrated man's ability to stay in space continuously for up to 14 days; Gemini 5 for 8 days, and two other missions for 4 days.

Controlled Reentry: Landing accuracies of within a few km of the aim point were demonstrated on every Gemini manned mission except Gemini 5.

Experiments: Every manned Gemini mission conducted many experiments. In total 43 experiments were conducted successfully.

Of the 14 Gemini mission attempts, 10 missions accomplished all of the primary mission objectives specified before the launch. The four unsuccessful missions, and the reasons why they could not accomplish all of their primary objectives, are as follows:

Mission	Reasons Mission was Unsuccessful
Gemini 6	The Agena target vehicle exploded. The Gemini 6 spacecraft was successfully rendezvoused with the Gemini 7 spacecraft later during the Gemini 6A mission.
Gemini 8	An orbital maneuvering thruster malfunctioned, which ruled out the primary EVA objective.
Gemini 9	An Atlas booster failure drove the Agena into the Atlantic and the Gemini 9 spacecraft was not launched until later during the Gemini 9A mission.
Gemini 9A	The shroud did not come loose from the Augmented Target Docking Adapter, precluding docking—a specified primary objective for the mission.

Gemini Launch Vehicles

The modified Titan-II missile used as the Gemini Launch Vehicle was 100 percent successful in Project Gemini (12 launches).

Gemini Target Vehicles

Six Agena target vehicles were launched and four were successfully placed in orbit, rendezvoused and docked with. The Augmented Target Docking Adapter, launched as a back-up target for the Gemini 9 spacecraft to rendezvous and dock with, functioned properly; however, the shroud failed to separate, thereby making docking impossible.

GEMINI AROUND THE MOON

During Project Gemini several proposals were put forward to extend the program to lunar missions—orbiting the Moon and even one landing proposal. None of them were implemented and they each received various levels of consideration and support. The most formal of these proposals, dated July 1965, was titled Rendezvous Concept for Circumlunar Flyby in 1967. An excerpt from its summary reads as follows:

> This study was made in collaboration with NASA Manned Space Flight Center and the McDonnell Aircraft Corporation to determine how a circumlunar flight could be made during 1967 using existing space hardware building blocks and operational techniques that will have been developed by 1967. Several approaches have been considered using the Titan-IIIC booster; e.g., a direct flight approach and an Earth orbit rendezvous. The most practical concept, which is detailed in this report, consists of launching a modified Gemini A capsule by a Gemini launch vehicle (GLV) to rendezvous with a stripped Titan-III transtage propulsion module containing a modified Agena target docking adapter (TDA). The propulsion module will be launched into orbit by a standard Titan-IIIC booster ...

> ... No system changes are necessary to the standard Titan-IIIC booster or the Gemini launch vehicle for this mission. The interface between the Gemini capsule and the stripped transtage propulsion module is designed to use the standard nine-wire umbilical connector being used in the present Gemini rendezvous program ...

It would seem that if Apollo plans had not already been so far along, Project Gemini perhaps possessed the ability to put men on the Moon.

One thing that most of the "Lunar Gemini" proposals had in common was the selection of Earth Orbit Rendezvous—two spacecraft meet in Earth orbit and their combined hardware and fuel are sufficient for a Moon mission. This was basically because Gemini-era launch vehicles couldn't lift enough mass to orbit with a single launch. The above proposal was for a lunar fly-by; lunar landing missions would be even farther outside of those launch capabilities. There were missions proposed to do a lunar fly-by or landing with a Gemini

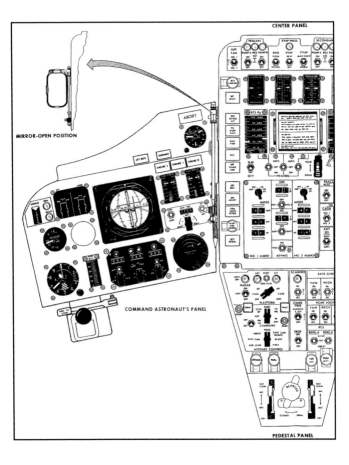

spacecraft and a Saturn IB or a Saturn V launch vehicle, but if you're going to do that, you might as well do Apollo. And at the time that these proposals were being presented Apollo was already well underway. It is an unfortunate commentary that today we don't have anything to match the Saturn V, and in fact US launch capability is not that far above that of the late 1960s.

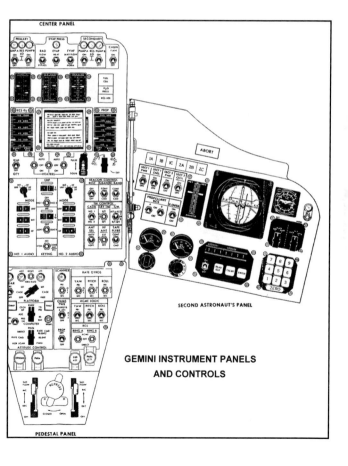

GEMINI INSTRUMENT PANELS
AND CONTROLS

GEMINI LAUNCH SEQUENCE

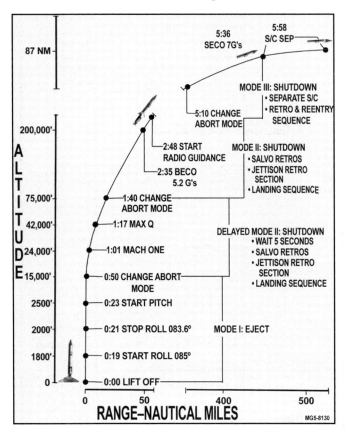

A typical Gemini mission launch sequence, including the possible launch abort modes, is shown above. The Titan-II was a two-stage booster and on a Gemini launch the second stage made it all the way into orbit. During the first orbit of the Gemini 7 mission the spacecraft performed orbital rendezvous and stationkeeping with the spent second stage.

GEMINI REENTRY SEQUENCE

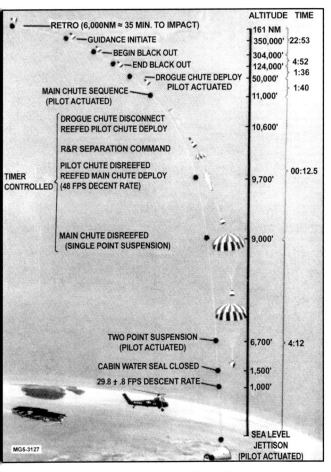

A typical Gemini mission reentry and landing sequence was executed almost entirely automatically, but with pilot backups for essential functions. The smaller drogue parachute was used to begin slowing the spacecraft's rate of descent in atmosphere and to help deploy the larger main chute.

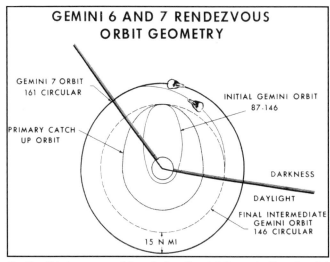

On the third apogee of its mission Gemini 6 was trailing Gemini 7 by about 300 km. At that point the rendezvous radar and onboard computer took over and brought Gemini 6 to Gemini 7's doorstep. The final, closest adjustments were made manually.

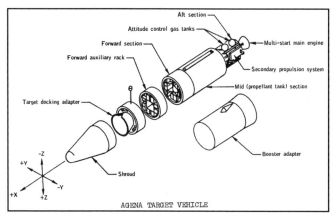

The Gemini Agena target vehicle was a modification of the U.S. Air Force Agena D upper stage. It acts as a separate stage of the Atlas/Agena launch vehicle, placing itself into orbit.

America's "sports car" spacecraft – the two-man Gemini.

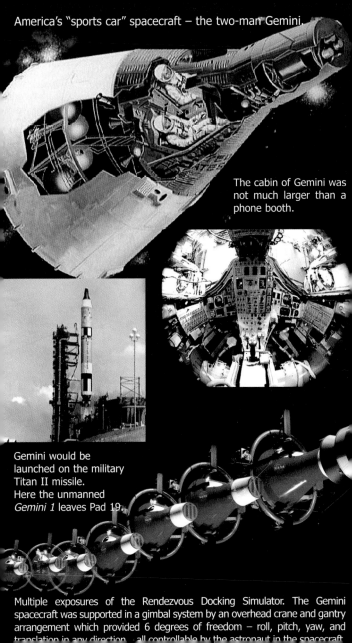

The cabin of Gemini was not much larger than a phone booth.

Gemini would be launched on the military Titan II missile. Here the unmanned *Gemini 1* leaves Pad 19.

Multiple exposures of the Rendezvous Docking Simulator. The Gemini spacecraft was supported in a gimbal system by an overhead crane and gantry arrangement which provided 6 degrees of freedom – roll, pitch, yaw, and translation in any direction – all controllable by the astronaut in the spacecraft.

Chosen to be first to fly in Gemini were Gus Grissom and John Young. They would pilot *Gemini 3*.

GRISSOM-YOUNG
GT-3
MOLLY BROWN

After flying *Gemini 3* astronauts John W. Young (left) and Virgil I. "Gus" Grissom took part in training exercises as the backup crew for the *Gemini* 6 mission, which featured the first rendezvous of two spacecraft in orbit

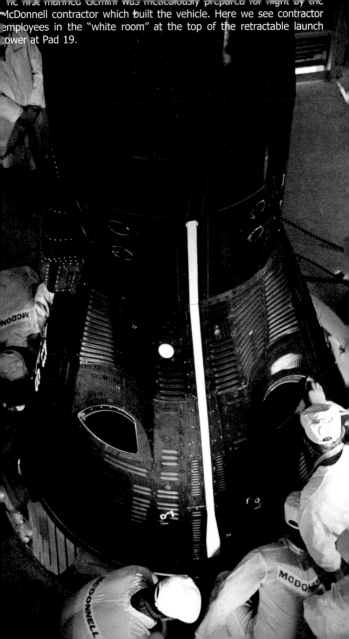

The first manned Gemini was meticulously prepared for flight by the McDonnell contractor which built the vehicle. Here we see contractor employees in the "white room" at the top of the retractable launch tower at Pad 19.

Gemini 3 leaps away from Pad 19 carrying Grissom and Young on March 23rd 1965.

View to the northwest over northern Mexico and southern California. The dark area in the middle is cultivated land of the Imperial Valley around the delta of the Colorado River.

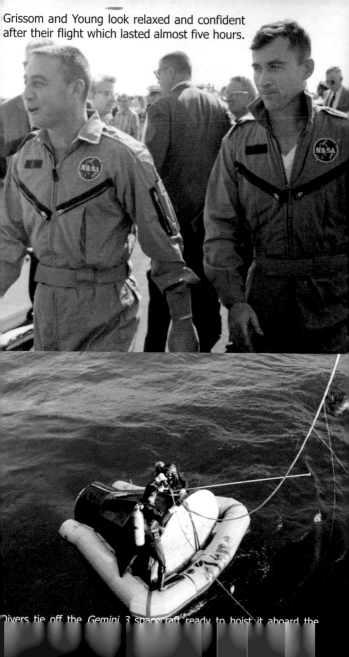

Grissom and Young look relaxed and confident after their flight which lasted almost five hours.

Divers tie off the *Gemini 3* spacecraft ready to hoist it aboard the

Next up was *Gemini 4* with Edward White (left) and James McDivitt.

GEMINI 4
FIRST
SPACEWALK
McDIVITT
WHITE

This shot shows how confined the inside of Gemini was. Here we see McDivitt and White preparing for departure.

Gemini 4 departing on June 3rd 1965.

White is seen holding his gas-powered maneuvering unit in his right hand. This allowed him a certain amount of control over his movements.

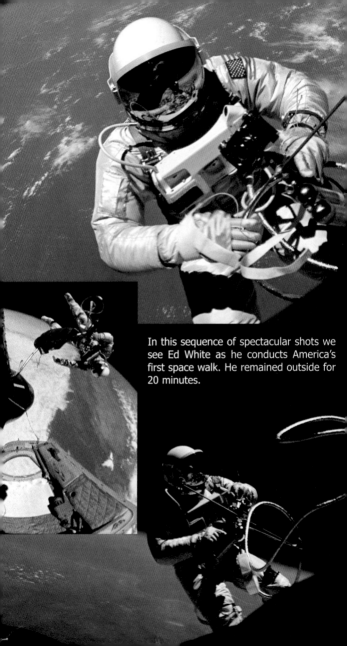

In this sequence of spectacular shots we see Ed White as he conducts America's first space walk. He remained outside for 20 minutes.

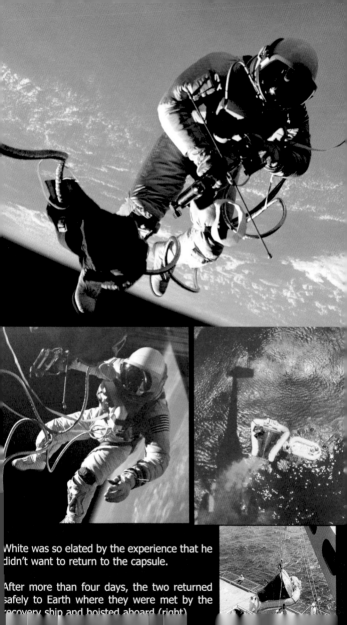

White was so elated by the experience that he didn't want to return to the capsule.

After more than four days, the two returned safely to Earth where they were met by the recovery ship and hoisted aboard (right).

Mercury veteran Gordon Cooper (right) teamed with Pete Conrad in *Gemini 5*. The two were both irrepressible jokers and made a memorable team. Here they are seen in training for recovery (below).

GEMINI 5
8 Days
or Bust
COOPER CONRAD

Cooper and Conrad walk towards their launch vehicle (top).
Here we see the white room and gantry as it retracts ready
for launch, August 21st 1965.

Gemini was controlled from the new Mission Control center. Note the screen at the front of the MCC which is used to track the progress of the Gemini spacecraft.

Conrad during the *Gemini 5* mission (below) and being hoisted aboard the recovery helicopter (left).

The recovery divers jump into the water to secure the vehicle after splashdown on August 29th 1965. The green dye in the water helps to locate the vehicle from the air (below and right).

Gemini 5 is hoisted aboard the recovery ship.

Conrad looks relaxed after almost eight days in space.

Conrad and Cooper
glad to be home.

Tom Stafford and Wally Schirra would fly *Gemini 6*.

The *Gemini 6* spacecraft (right) and Agena Target Vehicle (left) on the Boresight Range Tower for at the Kennedy Space Center to test the two spacecraft's docking capability. The Agena was designed to launch separately from Gemini and act as a target for astronauts in a Gemini spacecraft to rendezvous with. *Gemini 6* was slated to be the first mission to dock with an Agena, but a malfunction with the unmanned target resulted in new objectives calling for a one day rendezvous with *Gemini 7* in December, 1965.

Astronauts Walter M. Schirra Jr. (seated), command pilot, and Thomas P. Stafford, pilot, *Gemini 6* prime crew, go through suiting-up exercises in preparation for their flight. The suit technicians are James Garrepy (left) and Joe Schmitt.

The *Gemini 6A*, scheduled as a two-day mission, was launched December 15, 1965 from Pad 19, carrying astronauts Walter M. Schirra Jr., Command Pilot, and Thomas P. Stafford, Pilot. *Gemini 6* rendezvoused with *Gemini 7*, already orbiting the Earth.

Gemini 6 as seen by the crew of *Gemini 7*. This was the first true controlled rendezvous in space.

Another close approach to *Gemini 6.*

Schirra is clearly very pleased as he arrives on the carrier deck after over a day in space (below).

Assigned to the most gruelling mission of the Gemini program were two rookie astronauts, James Lovell (left) and Frank Borman. They flew in *Gemini 7* (below) and tested new space suits (right).

The *Gemini 7* was launched on December 4th 1965. The crew would rendezvous with *Gemini 6* and would then remain in space for the duration equal to a trip to the Moon.

BORMAN · LOVELL

VII

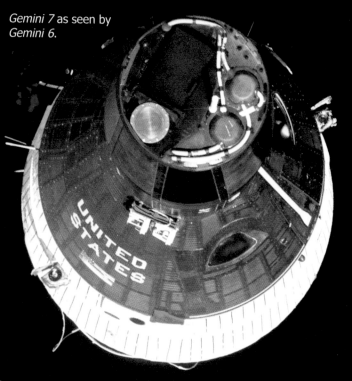

Gemini 7 as seen by *Gemini 6*.

UNITED STATES

Borman and Lovell spent almost two weeks inside the tiny Gemini capsule and found it difficult to walk when they arrived back on Earth.

Dave Scott and Neil Armstrong are seen here with a model of Gemini just before the flight of *Gemini 8*. Their backup crew was Pete Conrad and Dick Gordon (inset below).

Scott and Armstrong were to meet the Agena target vehicle seen on its Atlas booster at left, launched less than two hours before the Gemini on March 16 1966 (right).

Two spectacular views of the Agena target vehicle just prior to the ill-fated docking with *Gemini 8*. A thruster failure just after the first-ever successful docking caused the two vehicles to start tumbling and nearly caused a fatal accident.

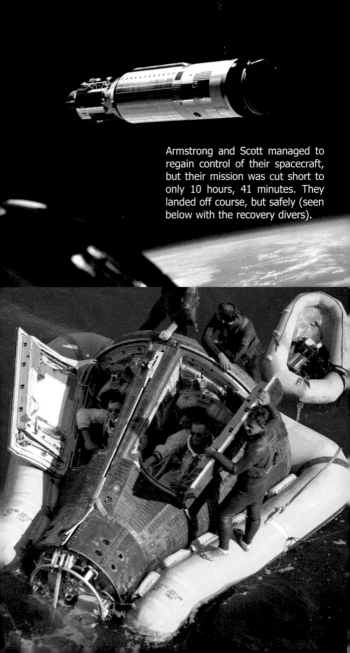

Armstrong and Scott managed to regain control of their spacecraft, but their mission was cut short to only 10 hours, 41 minutes. They landed off course, but safely (seen below with the recovery divers).

Tom Stafford (above left) would become the first to fly Gemini twice when he commanded *Gemini 9A*. Here he is seen with his pilot, Eugene Cernan. Below is the original prime crew, Elliot M. See Jr. (left), command pilot, and Charles A. Bassett II, pilot. Both were killed in a plane crash on February 28, 1966.

Astronaut Eugene A. Cernan, pilot of the *Gemini 9A* space flight, participates in extravehicular training under zero-gravity conditions aboard a KC-135 aircraft. Here, he is donning the Astronaut Maneuvering Unit (AMU) back pack after egressing a Gemini mock-up. The AMU back pack is mounted in the adapter equipment section of the mock-up. Cernan wears an Extravehicular Life Support System chest pack.

Astronauts Thomas P. Stafford (foreground), command pilot, and Eugene A. Cernan, pilot, walk up the ramp at Pad 19 during the *Gemini 9A* prelaunch countdown.

Astronauts Thomas P. Stafford, command pilot, and Eugene A. Cernan, walk away from Pad 19 (opposite) after the *Gemini 9A* mission was postponed. Failure of the Agena Target Vehicle to achieve orbit caused the postponement of the mission. The launch finally happened on June 3, 1966 (above).

Seen here is the Augmented Target Docking Adapter (ATDA) sent to meet *Gemini 9A*. One of the metal straps can be seen wrapped around the white shroud. This strap failed to separate and effectively stopped Stafford and Cernan from effecting a docking. Stafford said it looked like an angry alligator.

The successful splash down of *Gemini 9A* took place on June 6, 1966. Recovery vehicles were close enough to take this photograph.

The *Gemini 9A* spacecraft on display in a museum (below).

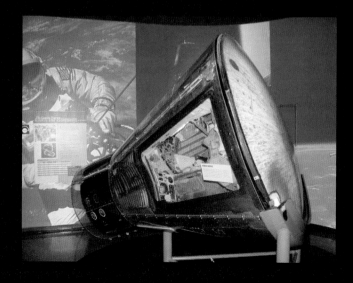

Time lapse photography shows the tower retraction and launch of the *Gemini 10* spacecraft on July 18, 1966.

John Young (left) returned to space for his second trip as commander of *Gemini 10*. His copilot was Michael Collins. Below they head up the ramp to the spacecraft in preparation for launch, July 18th 1966.

The *Gemini 10* spacecraft is successfully docked with the Agena Target Vehicle. The Agena display panel is clearly visible, as is the glow from Agena's primary propulsion system. A beautiful shot of Earth (below) was seen by Collins and Young.

The Agena target vehicle used by *Gemini 10* is seen above; splashdown (right) and Young heading up to the recovery helicopter (below).

Young holds forth on the pleasures of space travel with John Stonesifer (left).

Navy buddies Dick Gordon (left) and Pete Conrad flew the *Gemini 11* spacecraft seen here under inspection by McDonnell workers.

The Agena target vehicle targeted by *Gemini 11* was launched one orbit ahead of *Gemini 11* at 1.05 p.m. on September 12, 1966.

At 2.42 p.m. on September 12, 1966, *Gemini 11* soared aloft from Pad 19 carrying Dick Gordon and Pete Conrad. Once again, the copilot, in this case Gordon, was to take an extensive space walk (inset above).

The Agena Target Docking Vehicle is tethered to the *Gemini 11* spacecraft during its 31st revolution of the Earth. Below is a spectacular shot of the southern tip of India and Sri Lanka off to the right.

The famous "space cowboy" shot of Dick Gordon straddling the Agena.

Gemini 11 is back home after nearly three days having flown farther from Earth than any other manned spacecraft to date.

Gemini 11 and its attendant Agena Target vehicle both launched on November 11, 1966 within fifteen minutes of each other.

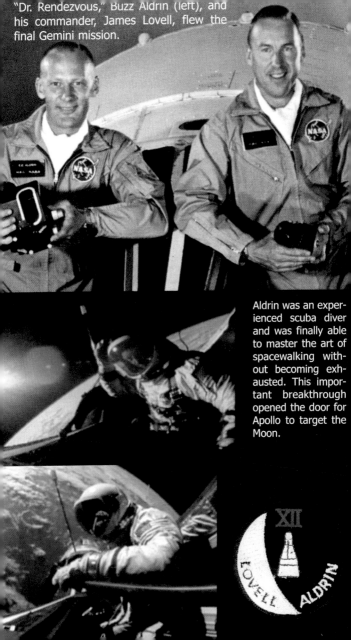

"Dr. Rendezvous," Buzz Aldrin (left), and his commander, James Lovell, flew the final Gemini mission.

Aldrin was an experienced scuba diver and was finally able to master the art of spacewalking without becoming exhausted. This important breakthrough opened the door for Apollo to target the Moon.

XII

LOVELL ALDRIN

Aldrin stayed outside for a total of five and a half hours.

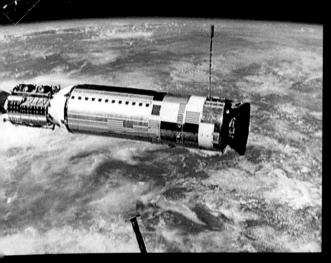

Lovell and Aldrin successfully rendezvoused and docked with the Agena (above). They then remained tethered to it to conduct experiments (below)

After nearly four days, *Gemini 12* returned exactly on target, as can be seen by the proximity of this rescue helicopter (right). This was the end of the tremendously successful Gemini program.